Essays in AI:

Automation, Technology and

the Future of 9-5 Work

By Joshua Krook

Table of Contents

Chapter 1:..1
On the Future of Pointless Jobs, the 9-5 Work Day and
the Robot Revolution...1

Chapter 2:..13
Towards a Project-Based Economy: Ai and the Future
of Work..13

Chapter 3:..18
Pokemon Go, Augmented Reality and the Future of
Technology...18

Chapter 4:..24
Automating Job Applications – Should a Robot Apply
for a Job on Your Behalf?..24

Chapter 5:..29
The Right to Personality in the Workplace
(Dismantling the Private and Public Self......................29

Chapter 6:..37
On Advertising and the Loss of Free Will.....................37

Chapter 7:..46
Modern Life: A Consciously Celebrated Materialism. 46

Chapter 8:..52
Lifetime Employment in Japan: Casual Work, Part-
Time Work and Women Under Equal Opportunity Law
....52

Chapter 9:..99
20 Years on From Gangland: We've Still Got a Youth
Culture in Crisis...99

Chapter 10:...106
Degrees of Separation: Companies Shed Degree
Requirements to Promote Merit Over Qualifications
....106

Afterword...110

Chapter 1:

On the Future of Pointless Jobs, the 9-5 Work Day
and the Robot Revolution

When we see the world through the lens of how things currently are, it is almost impossible to imagine things differently without in some way imparting some of the existing dynamics into the imagined future.

Ask a random person on the street today why they work a nine to five job, and they will give you a variety of personal answers. Insist: *No, why do you work nine to five in particular, why that number of hours?* And they will give you some derivation of "that's just the way the world works." Eight hours is average, and most people work the average number of hours because that's what average means.

But is it true that "that's just the way the world works"?

In 1834, Connecticut, what was average meant something completely different to today. Employers in large factories had discovered at the beginning of the Industrial Revolution that factory output was maximised —and more goods were produced- if employees worked up to 20 hour days, 6 days a week.

It was average for employees in factories to do factory work dawn to dusk, in what we now consider unbearable conditions.

We can imagine a man (actually it was mainly kids) on a factory floor in 1834 thinking over his long working hours and saying to himself or a friend, "well, I guess that's just the way the world works."

And yet such a man (or kid) would be proven wrong a mere year later.

In 1835, a worker's strike in Patterson, New Jersey textile mills sparked widespread union protests all along the East Coast of America that caused a change to the number of working hours in factories from an unregulated infinity, to a regulated 10-hour working day.

In 1914, Henry Ford unilaterally reduced his own employee's workday from 10 to 8 hours, from a 6-day week to a 5-day week. Ford stated at the time: *"It is high time to rid ourselves of the notion that leisure for workmen is either 'lost time' or a class privilege."*

Following the change, productivity at Ford Motors skyrocketed, confirming later managerial theory that being decent to employees boosts overall productivity.

Largely inspired by the story of boosted productivity, other companies followed suit.

The Ford Motors story produced the current employment system as we understand it today and the nine-to-five, five days a week norm that remains the national average.

It is important to remind ourselves from these historical changes and trends that the world does not "work" in any particular way whatsoever. Rather, the way the world 'works' or does not work is subject to changes brought about by human beings.

This is particularly the case when it comes to systems that were designed, built and maintained by humans to begin with — such as the employment system.

Our Current System Dates Back to 1914; So What's Next?

It's sometimes strange to think how *little* the work hours of the population have changed in the over a 110 years since 1914.

While part-time work, shift work, casual and flexible work practices have all become common, it is still average, and indeed expected, to have and to hold a 9-5 job.

In an article in *Strike Magazine*, anthropologist David Graeber asks the crucial question: why?

> *In the year 1930, John Maynard Keynes predicted that technology would have advanced sufficiently by century's end that countries like Great Britain or the United States would achieve a 15-hour work week. There's every reason to believe he was right. In technological terms, we are quite capable of this.*

> *And yet it didn't happen.*

> *Instead, technology has been marshalled, if anything, to figure out ways to make us all work more. In order to achieve this, jobs have had to be created that are, effectively, pointless.*

Instead of technology being used to free us from a forty-hour week, we have somehow gone the other way, and used technology to reinforce the system that Ford originally established in 1914.

Logically, this makes little sense.

At a basic intuitive level, computers speed up the rate at which humans can work, meaning that, by and large, we should be accomplishing more in less time; begging the question, why are we all spending the exact same amount of time at work? This is particularly perplexing regarding the effect automation has on the basic necessities of human existence like: food, clothing and housing, and the increasingly expanding human population.

One could imagine a 3D printing revolution where food and clothing are available in every household at the press of a button; effectively making 'working to make ends meet' an outdated anachronism.

But Graeber goes further to suggest that we're kind of already there:

> *Over the course of the last century, the number of workers employed as domestic servants, in industry, and in the farm sector has collapsed dramatically. At the same time, "professional, managerial, clerical, sales, and service workers" tripled, growing "from one-quarter to three-quarters of total employment."*
>
> *In other words, productive jobs have, just as predicted, been largely automated away. But*

rather than allowing a massive reduction of working hours ... we have seen the ballooning not even so much of the "service" sector as of the administrative sector, up to and including the creation of whole new industries like financial services or telemarketing.

These are what I propose to call "bullshit jobs."

With the automation of agriculture and farming, it was expected that the leisure time of the majority of the population would dramatically increase.

This was not simply speculation, but a product of historical consideration.

When looking back at feudal times we can gather that the aristocracy and noble class had ample free time, largely due to the diligent work of serfs below them; producing the food and sustenance necessary for their continued survival.

Now that we have machines (and, albeit, a very small percentage of the population controlling machines) producing our food, agriculture and clothing for us, it begs the question; where's the free time we were meant to get?

Graeber suggests that our free time has gone into what he calls "bullshit jobs," plumped up positions that do nothing for society except produce paperwork and move money between people, apparently for no reason whatsoever.

Huge swathes of people, in Europe and North America in particular, spend their entire working lives performing tasks they secretly believe do not really need to be performed.

Consider telemarketers being replaced by machines, or bell boys being replaced by automatic rotating doors, or even sales and marketing staff being replaced entirely (no such staff used to exist, and yet now entire industries are devoted to marketing, sales and advertising – why?).

The sheer diversity of products we are exposed to likewise goes beyond what is sane or sensible. At a certain point materialism bombards us with so many choices and so many items that it must make us pause and think, should we even be materialistic at all?

The original Ford motorcar was only in black – now there's no point arguing for a black and white society today, but if we took the edges off the colour palette… If we got rid of some of the extraneous variations on items so as to give the entire country an extra day off per week – if we cut up extraneous industries and marketing staff and pointless frivolities of employment, then maybe we may be onto something here.

It's likewise difficult to think of menial tasks in particular, such as cleaning, that can't *already* be automated, or soon shall be, to the point where humans no longer need to do them.

We've all seen advertisements for things like the KL-310 vacuum cleaner able to clean a floor by itself without any human operator.

And yet this is merely the beginning of such a phenomenon.

Almost all artificial intelligence experts say that we're verging on a breakthrough on the subject of robotic artificial life by about 2050.

The robotic revolution, and everything that means for human employment, is about to begin.

NB: It is important to note the likely increased automation of industries even before a complete artificial intelligence is created. In the next 40 years things like domestic housework in particular are likely to undergo significant changes into automation and those manual industries will dissolve, likely as is historically the case with such things, into sales and marketing. God knows why.

The Robotic (Post-Industrial) Revolution:

There is something very curious about politicians constantly obsessing over people getting jobs in the light of the oncoming Robot Revolution.

Now you might think I'm crazy for believing in such things, but then you will have to call the likes of Stephen Hawking crazy too, which is a much, much more difficult task.

There are already articles on the web asking: *"What will happen when Robots Take our Jobs?"* The idea is that an oncoming robotic revolution is coming whether we like it or not.

And with it, the capacity of robots to do the jobs typically reserved for humans – including high-end, white-collar professional work. The latest robotic innovations out of Japan can play ping pong ("and even decide to take it easy on opponents by missing a few hits"), use sign language to "talk" to humans and "mimic simple greetings." This is only the beginning.

Despite almost every single instinct of intuition in my body saying that robots will make our lives easier, which is what we've been taught (using examples like the washing machine in the 1950s) —by freeing up our time and allowing us to work on things that aren't menial, boring office jobs— we have to look to history here and realise that that seems like an unlikely outcome. History has a few examples where this is true, but on the whole it has gone the other way, and this time round...

It may even go the other way.

Despite huge, widespread automation since 1914, we still have the exact same 8-hour workday as workers in Ford did back then. Our work hours are static, unchangeable even, amidst growing and growing automation.

Why do we think getting robots will make any difference here, if other kinds of automation haven't? Why do we think robots will free up our time, when factory machines and agricultural automation did nothing of the sort?

In fact, scientists are predicting that robots will make our lives even worse. Their predictions are based on the idea that our employment systems will in fact, as has occurred over the last century, remain exactly the same.

Our unchanging systems will, in effect, be the cause of our greatest decline – by obsessing over keeping things as 'the way the world works' – the world will change dramatically from under us and ruin all our so-called workings.

Oxford Philosopher Nick Bostrom recently told *New Republic*:

> *The subsistence level for digital minds would be a lot lower than for biological minds. Biological humans need to have houses—we need to eat, we need to transport ourselves. Digital minds could earn, like, a penny an hour. The wage level would fall; humans could then no longer earn a wage income. It looks very questionable, in this free-for-all competitive world, that we would find a niche for our small, stupid, obsolete minds.*

The prediction is that our current working system would continue exactly as it has for the last century. Except, instead of humans outcompeting each other, robots would outcompete humans. Employers would pay the lowest wages possible, relying mainly on robots, and then instead of redistributing this into freeing up humanity's time, or distributing the money to humans in general– humanity will simply lose out. Now that's bloody well grim isn't it?

How do we prevent that kind of future?

Change the employment system we have now.

*

Failure to ask questions about the future of our society is an implicit endorsement of the direction our society is already heading in.

Chapter 2:

Towards a Project-Based Economy: Ai and the Future of Work

An AI lawyer is doing legal research in the US, a robot is laying bricks in Japan and a robot just passed a visual Turing testat MIT. The question is no longer whether automation will occur but how long we have to control its introduction and the future of human work before it does.

While the media continues to argue over what the utopian future will look like when robots do our laundry and dishes, a much bleaker picture is emerging of humans working unregulated working hours in labour camps under crippling working conditions to 'compete' with their robot colleagues in the workplace. Robots are estimated to potentially work for just a penny a day, making them much tougher competition than the supposed threat of low wage immigrants.

In this essay, I intend to argue against the popular misconception that robots and AI will lead us towards an unprecedented revolution in individual freedoms.

Since the Second World War the story has gone in the opposite direction, as technology and automation has been used to make workers work longer hours. The anthropologist David Graeber points this out.

Despite significant changes in technology in the last few decades there has been almost no reduction in working hours. Instead, our economy has gone in the opposite direction, creating 'filler jobs' or 'bullshit jobs' as he calls them, to fill in the vacuum of dying automated industries. Where automation was meant to free up people to do what they like with their time, it has instead been used to make us work longer.

The average full time working hours in Australia remain at forty hours a week, with around seventy percent of the economy working a full time job. The majority of the population do not work in food, clothing or shelter but in ancillary "luxury" industries like advertising. Jobs in advertising, marketing, social media and so on have seemingly been invented for the sake of keeping us working longer hours, as opposed to any direct survival-based output. They hardly have an evolutionary imperative, but rather an imperative of social continuance – of people being retrained and re-skilled to keep them working.

This process of inventing jobs to keep people working is at the heart of why AI and automation may not free up our lives in the future, even with the resurgence of robotics. Whenever automation has been close to granting people individual freedom – a new industry has risen up to keep people working the same hours, indefinitely.

Our politicians are already talking about AI in the same way as they discussed previous introductions of automation. They talk primarily of training young workers for "jobs that don't exist yet," for an "uncertain future" and so on.

When they admit that automation will eliminate certain white collar industries, they call for a reskilling of workers to fit into "new industries," yet to be created. This is coded language for the idea that working hours will never be decreased regardless of how much the economy gets automated.

Instead of reducing working hours new industries will be created to force humans into servicing robots and AI machinery. The recent focus on "STEM skills" and "coding" in this context is particularly pernicious. There is a reason why 'coding' is such a buzz word, and it has nothing to do with politicians liking to code. Facing the prospects of a 40% unemployment rate, the government can do little but force young people into narrower and narrower specializations with regards to robotics and technology, in an attempt to prepare them to create "future industries" using "STEM skills". Indeed, a recent Oxford study predicted that around 35% of current jobs are at risk of automation over the next ten to fifteen years.[1] The remainder will be technocratic positions to manage the fallout from automative work.

If this all seems farfetched, a recent example should give us reason to pause. Perhaps the most famous example of automative technology in the last century, the computer. Both the computer and the internet were once heralded as the main harbinger of individual freedom. They were expected to lower working hours. Computers were expected to free up our time while drastically increasing our productivity.

In reality, productivity increased but working hours remained static – wages flatlined and living conditions for the working poor declined dramatically. Blue collar workers in particular suffered declining standards of living, marking decades of decline.

Despite computational technology, or perhaps because of it, we are working longer hours than ever before. Employers can now contact their employees outside traditional work hours– by email, text, social media and so on. Instead of freeing up our time to spend on projects of passion, automation over the last two decades has been used to make us work 24/7. Erasing the very concept of a 'private life'. The problem is so bad that the French government recently banned employers from emailing their staff after office hours.

The story wasn't meant to go this way.

In 1933 the economist John Maynard Keynes famously predicted that in the 2030s humans would have a 15-hour workweek due to massive advances in technological innovation and change. Keynes prophesised that people in the 2030s would ask not what to do with one's work but what to do with one's life in general. Human purpose would replace money as the central question of value in society leading to a new Renaissance of individual liberty and purpose.

Instead of preparing young people for "jobs that don't exist yet," political leaders should learn the lessons of history from the creation of the computer. Instead of working the same or longer working hours, AI should be utilised to replace human labour entirely – allowing for Keynes' 15 hour working week to be realized.

While many are suggesting a basic wage as a possible silver bullet for the problem, it is clear that at a minimum a complete re-think of the nature of work is in order.

Chapter 3:

Pokemon Go, Augmented Reality and the Future of Technology

In 1974, the philosopher Robert Nozick came up with the idea of an 'experience machine', a thought experiment which involved the following situation:

> Suppose there was an experience machine that would give you any experience you desired. Super-duper neuropsychologists could stimulate your brain so that you would think and feel like you were writing a great novel, or making a friend, or reading an interesting book. All the time you would be floating in a tank, with electrodes attached to your brain.

The experience machine would give you the *experience* of having done something, without doing anything at all.

If the idea sounds familiar it's because the idea was made into a film; the *Matrix* franchise.

In the Matrix, humans are literally plugged into a virtual world (*The Matrix*) while their real bodies float in tube-like containers.

In his thought experiment, Nozick asked the following:

> *Would there be any difference between the experiences we have in an experience machine (e.g. The Matrix), as compared to the experiences we have in real life?*

It's a great question, and for his sake, Nozick ends up concluding that no, a simulation cannot beat reality.

Modern Day 'Experience Machines'

But Nozick answered that in a thought experiment. A hypothetical. Modern society is advancing to a time in which we'll be able to answer Nozick's question directly.

The latest advancements in VR (Virtual Reality), along with digital devices that augment our reality (Google Glass, Smart Watches and so on), bring us closer than ever to a kind of hyper-reality, where we can experience reality through machinery.

Instead of living in the real world, we live in a technological world, an imagined world, or an

augmented world, 'plugged in' to different devices that act as our own experience machines, to give us simulated experiences that simulate or emulate everyday life.

On one end of the spectrum is Virtual Reality, which allows people to walk 'inside of' imaginary worlds through the use of a headset (funky goggles that basically act like TV screens on your eyes), and integrated video gaming technology (allowing you to interact with the world around you that you are 'walking' through).

Augmented Reality:

On the other end of the spectrum is a thing called *augmented reality*. Instead of wearing goggles and entering an imagined world, augmented reality superimposes an imagined world onto our own.

The best example of this is in a new video game called *Pokémon Go*. Here's how the press release describes it:

With Pokémon GO, you'll discover Pokémon in a whole new world—your own!

Pokemon Go is a form of augmented reality, where players catch Pokemon in the real world, through the use of a wristwatch-type-thing and GPS tracking technology.

Basically, the game allows you to catch Pokemon outside, on the street.

In my estimation, *Pokemon Go* is the closest thing to Nozick's 'experience machine', a form of simulated reality.

Why Should We Care?

What happens when this kind of technology is used to simulate real life experiences, as opposed to just fictional worlds?

Imagine a world where, instead of playing soccer, you play virtual soccer; or instead of dating, you date a virtual person; or instead of traveling, you travel 'virtually' to different countries.

There's something insidious in these 'progressions' or 'advancements'.

In the wrong hands, augmented reality could be used to suppress human movement and actual, authentic experiences in everyday life in exchange for simulated experiences. Bosses could have us 'go on holiday' while not leaving our chair in the office.

In fact, we could live our entire lives from one room.

Why go to Europe when you can go to virtual mars?

The Great Disconnect:

Increasingly, the world is relying on self-medication to deal with systematic problems in our lives.

People drink excessively on the weekend to escape the job they hate in the week; they travel one month in a year to escape the 11 months they dislike; they go on Tinder dates to avoid sustained relationships.

Augmented and virtual reality are the next step in this kind of evasive-living.

Instead of actually solving the problems in our lives, we simulate perfect lives in fictional worlds. In a way, this pacifies us into a kind of entrenched inactivity that becomes harder and harder to shake off over time.

Technology becomes less of our friend and more of our master.

The lines between fantasy and reality become blurred beyond recognition.

And we live in a pseudo-reality, very different to our own.

Chapter 4:

Automating Job Applications – Should a Robot Apply for a Job on Your Behalf?

An estimated 90% of large companies are using automated software to read and respond to resumes. From tracking software that reads a resume looking for keywords, to automated emails, phone interviews and skills testing – companies are doing less manual processing than ever before.

If companies are using automated software in hiring, then job applicants should be able to use the same technology.

In an increasingly competitive job market, with some advertisements attracting hundreds of thousands of applications, there is less chance that the time spent on putting together an application will be rewarded. As the number of applicants increases, the success of any one applicant decreases. In other words, job applications have become a numbers game.

In the hands of applicants, automated software would drastically cut down the time taken to apply for the requisite number of jobs to secure a position. Automating the process would reduce time spent reading job ads, uploading resumes and signing on to

various job sites under the current, woefully ineffi-
cient system.

What a job application robot looks like

Several programmers have already attempted to create a job application robot. This type of online robot,
or "bot", is a programmed piece of software that per-
forms a variety of functions on behalf of the applicant.

    The most common type of such software has three
core components. The bot runs a trawling algor-
ithm that searches the internet for relevant job ads
and compiles them into a list. It then uses an email
template to create a cover letter, using keywords from
the advertisement and job description to tailor a resume for the position.

    To cap it off, the bot sends an automated email
applying for the job on your behalf to the relevant human resources manager, making sure to get their
name right. In effect, this is a total automation of the
entire job application process, from start to finish.

When will the robot wars start?

Despite the obvious appeal of fully automated job application bots to consumers, uptake has been relatively slow. Most of the automated bots thus far have been created by independent hobbyists, often as one-off experimental projects.

There have only been a few open source tools (that can be used by others to develop software) that cover part of the process. Consumers are also wary of using application bots, when human resources managers continue to insist that they manually spend time personalising each application.

Ironically, one of the biggest advantages of a bot-based application system is that it can come across as more personal than a human. If resumes are read initially by tracking software looking for particular keywords from job ads ("individualised" keywords), it makes sense to use bots that automatically include those key words in an application. A bot's application in this way can appear more "tailored" to each job ad than if written manually by a human being, who might overlook the importance of focusing on keywords.

In two test case scenarios, bots were found to outperform the manual applications of their human inventors. One programmer, Robert Coombs, found that his bot had an increased response rate of 11%compared with manual applications, out of a test of 538 jobs. Another programmer, Benjamin Derville, had a 50% response rate for 50 jobsusing a bot. It was the best response rate he had ever achieved when applying for jobs.

Despite this measured success, both program-mers warn that online applications are inherently rigged. With the rise of networking, nepotismand cronyism, companies are increasingly ignoring exter-nal applicants altogether.

Even though full automation might not have taken off just yet, partial automation of job applications has become increasingly common. Job applicant advice books now commonly include a section on automating one or two aspects of the job search.

Most commonly, these books will include the idea of "mail merging". This involves manually collecting data about various jobs in a spreadsheet, including job title, employer title, company name and so on. Data are then "merged" into several new cover letters in Microsoft Word, automatically.

In this way, form data is used to fill in hundreds of cover letters in a matter of minutes, rather than hours. On the institutional side, companies like LinkedIn are making it easier than ever for applicants to fill in their application data automatically from their existing profile, rather than starting from scratch in each application.

Considering the time saved, the efficiency gained, and the current asymmetry of automated technology, it seems clear that job applicants should be given access to automated job application bots. This would not only save everyone time, but would also drastically reduce the amount of human error in the system – leading to an employment system that uses new technology to free up our lives.

This article was originally published on *The Conversation*:

Read the original here:

https://theconversation.com/resume-robot-wars-how-employees-could-match-employers-use-of-tech-in-job-applications-80615

Chapter 5:

The Right to Personality in the Workplace (Dismantling the Private and Public Self

The ultimate aim of the employment system is not to *create* a self. This is a long-forgotten aim in some distant, lost century.

The aim now is to package the self, to create the illusion of a self in such a way that that self is appetising to those willing to try it out. This regime maintains a rigorous self-regulatory aspect: personal beliefs, political beliefs or criticisms are to be stored away in some *private* self, only to come out on weekends.

A cognitive dissonance exists between the public and private, to the point where we feel an innate sense of inauthentic vibes emerging from those around us. Social discussion becomes a cheap relay of pre-packaged emotional responses, conventions and platitudes that seek not to provoke but to quell provocation, not to excite but to dampen excitement, not to bewilder but to normalize interaction.

In the employment system we take on the role of an obedient, subservient member of a team.

> *"In the course of playing social roles, peo-*
> *ple often come to internalize role-relevant per-*

*sonal characteristics. They come to see them-
selves as possessing the qualities suggested
by the roles they play".[1]*

This is changing, with management theory revealing the benefits of democratic teamwork, devil's advocacy in the workplace and the general positive results that emerge from negative critique. But basic social convention and interaction often stays the same for long after these new 'techniques' are introduced.

When J.D. Salinger wrote *Catcher in the Rye* in 1951, he spoke of the "phony" characteristics of people in his generation, in some respects he was talking about the disconnect between private and public self. Holden Caulfield himself *presents* well, adults praise him (consider the nuns and his aging professor) but he does not reveal who he is. His private self is a lot more hostile than his public self, to the extent that no one can assist him with his troubles – he feels alone and isolated. The book resonates with teenagers for precisely this reason. Salinger was hailed as the "voice of his generation." But this praise remains incredibly ironic: he spent almost the entirety of his book criticising his generation. If he was the voice, where was the generation who followed his words with action? People still went on lying in interviews, dressing up who they were to impress others, putting on fake masks in different social settings depending on what was required of them.

These were the same people who said the book "resonated" with them.

The danger in creating multiple selves has long been understood in psychology. Roy Baumeister says:

> "*the concept of the self loses its meaning if a person has multiple selves...the essence of self involves integration of diverse experiences into a unity. In short, unity is one of the defining features of selfhood and identity.*"

If this is true, how can someone even *form* a self within employment, if they are required to split their personality in two? The distinction between private and public selves has grown starker and starker as the years go by.

The dangers are more prevalent, because, "when we publicly state an opinion or behave in a given fashion, we are expected to be the person we claim to be".[2] This issue raises a distinct possibility of cognitive dissonance. If we present only part of ourselves

during our working lives, and then our colleagues *expect* us to be that person always – we are
setting ourselves up to be oversimplified, and setting
up them to be deceived.

It is becoming increasingly common for people to
delete their social media accounts in the lead up to a
round of interviews. This is exactly what I'm talking
about. People are deleting their personalities to fit a
proscribed job role – even when their personalities
have nothing inherently offensive (or criminal, for instance) about them. Even when they are in almost
every way: completely normal.

I asked someone who deleted their personality
(both metaphorically through social media and in reality in the interview) why they did so. They answered:
"Once I went to a job interview where they printed out
my entire Twitter feed and went through it, comment
by comment, asking me about every single Tweet.
What I meant, what my views were, why I said what I
said? In sum, questions that were completely irrel-
evant to whether I could do the job or not."

People are so scared now of revealing their personal opinions on *anything* to their boss, be it religion,
politics or otherwise, that they are willing to delete
themselves to avoid the revelation. If they do reveal
anything about themselves they run the risk of getting

fired due to new standardized Australian contracts that ban people for talking about their work on social media: even if such discussion is in the public interest (consider corruption; sexism; racism or discrimination as examples, where public letters have done wonders for ending terrible practices).

The reality is that meaning comes foremost from self-definition and individualisation; the idea that we can gain meaning in conformity is anathema, mentally harmful and a damaging indictment on the level of doublethink our society has come to endorse.

We are what we create, and in our rush to create a self before we're ready to do so –and I speak here of the tens if not hundreds of people I know unsure of what to do but still proclaiming their love of a profession in interviews- leaves us unfulfilled because the self we are presenting is not who we actually are. This trend doesn't so much indicate a disconnection *of* the self; it indicates a disconnection *from* the self. What we are doing is separating ourselves from who we are for the tacit benefits that this process entails. We have locked up our individuality into the tightly confined boxes of our houses and expect that we will still be able to sustain meaning, fulfilment and self-esteem (from individual growth) outside of these boxes, even though honesty, frankness and free discussion are essential elements of individual growth. In other words, we are setting ourselves up for failure.

Worse than this is the persistent frequency with which this issue has been raised, time and time again, and consistently ignored, often provoking a vicious public backlash. Authors like Salinger, Bret Ellis, McInerney and David Leavitt were praised in their times as "voices of their generation". But this praise was consistently met by a furious backlash against 'whining' and 'complaining,' also known as a backlash against *caring* and *solving* problems. The antithesis to a first world problem is a first world solution.

The gradual resolution that society has come to is an **intermittent** celebration of the dissenting voice. If too honest, too painful, to grow up with the realization that some of our most fundamental systems are structurally and fundamentally incorrect – then we later discard that old realisation. Inspiring authors get condemned to the category of "youth" and get proscribed as young texts aimed at "rebellious teenagers". Instead of acknowledging criticism as valid, criticism gets acknowledged as "young". Salinger and others get lumped in this box, ready to head out to the garbage collector.

The implication is that by the time someone reaches adulthood they should stop criticising systems at all – rebellion is for the youth. But the youth – despite popular mantra – have the least power to change anything: they have none of the levers of "50 years experience in X" that society so heavily endorses. The political class has moved beyond the fundamental questions of improving life, improving work and disimproving people's ability to fake who they are, instead going on a tangential crusade about 'workers rights,' 'jobs' and dental.

Eight out of ten people hate their jobs. 64% of Australians are *actively* looking for new work. Free dental will not solve this: radical shifts in the way we practice our social interactions and the way we find meaning in our lives might.

But in the end, as much as we may condemn everyone else – we must ultimately take responsibility for how we are largely to blame. If there really are people who are the "voice of our generation," then they remain powerless and frankly, pointless, if everyone else in the generation remains silent.

The ability to accept something like *Catcher in the Rye* as a great book one day, and lie in an interview the next, displays our inability to reconcile our values with our present-day reality.

To do so is quite simple: stop applying for jobs you don't want. Only apply when you actually want something; be honest, care enough to tell an interviewer what you actually think.

The solution to being "phony," as Salinger called it,

is being real.

—-

[1] McCall & Simmons, 1966; Sarbin & Allen, 1968; Stryker & Statham, 1985

[2] (Goffman, 1959)

Chapter 6:

On Advertising and the Loss of Free Will

In his treatise on free will, philosopher Sam Harris claims that if an act is formulated in our subconscious then that act cannot be said to be *willed* by us. A reflex action to catch a ball, an instinctual action to blink in heavy sunlight or the act of breathing are all acts that are not consciously chosen by us, but automatic, reflexive or spontaneous actions.

Something that is automatic is not *willed* or decided upon, but, to a large extent, predetermined by the factors that initially create it. The blink is created by the sunlight, for example. And, for the same reason, we often cannot control these acts.

Subconscious decisions extend beyond reflex actions to decisions in everyday life.

The most famous example is the purchase of everyday consumer products, as influenced by the power of advertising.

To put it simply: Marketing is the art of selling something you didn't make, to someone who doesn't want it, for a price disproportionate to its worth.

In the 1950s and 1960s, the field of advertising had an influx of psychologists working around the clock to determine what *subconscious* motivators determined human action. The old paradigm of one kind of consumer motivated by *why* a product was a good purchase, was quickly subsumed by the idea that there were many *types* of consumers , all motivated by *subconscious* desires.

Instead of selling cars, advertisers began to sell male cars and female cars; old-people cars and young-people cars and so on, using the presumed personality traits of these distinctive demographic groups.

The biggest revelation however, came with the idea that:

> *"Human behaviour [was not a result of conscious thought but] a result of unconscious efforts to control inner drives and instincts motivated by petty emotions, sexual desire and anxiety"* – [1] .

Humans were not as disciplined and rational as first thought. But were instead irrationally motivated by their wants and needs (even if those needs were unknown to them at the time).

The Risks:

Our subconscious desires, by nature of being sub-conscious, are subject to a particular kind of psychological manipulation by advertisers. The most famous example is through the use of sexual imagery.

The epithet "Sex Sells" is based on the idea that sexual imagery in advertising often correlates to an increase in the sale of a product. (This is only partially true – as recent research finds that sex only sells if the product itself is considered 'sexy').

Sexual imagery in advertising does however, light up specific regions of the brain that are linked to an increase in financial risk taking, and so subconsciously compels us to take risks we may otherwise not have taken, towards, for instance, purchasing an advertised product.

Inspired by neurotechnological breakthroughs, the advertising industry has turned towards the use of brain scanning technology to determine the subconscious desires of consumers.

A recent book called Buyology, collected findings from numerous studies in neuromarketing, concluding that: subliminal advertising works (and bypasses conscious rational thought or conscious objections), emotional reactions to products can be implanted into the subconscious mind, and increased brain activity during an advert may be linked to an increased desire to buy products.

The fact that the advertising industry today is relying firstly on brain scanning technology is alarming. It is particularly worrying that these technologies are uncovering our subconscious motivations and desires (unknown to us), and revealing ways to shift these desires.

As Sam Harris suggests: if an act is formulated in our subconscious then that act cannot be said to be *willed* by us.

If an act, to purchase a product, is formulated first in our subconscious (as determined by our subconscious needs and desires, as shifted by advertising), and then concluded upon in our conscious mind (when we decide actively to buy a product based on said subconscious desires), then have we retained our free will?

What is worrying here is that our subconscious desires and needs, largely unknown to us, are being manipulated. Worse still, they are being manipulated by subliminal messaging, or carefully planted psychological marketing tricks.

It is not necessarily the case that advertisers are compelling consumers to buy particular products, but rather that they are compelling the subconscious needs and desires of consumers to shift towards a state that is best suited to buy said products.

Advertisers play on emotions of nostalgia (targeting Baby Boomers with music from their twenties) or play on the emotions of love and compassion (through scenes of romance or small children).

But the most prominent and most successful emotional tool that they utilise to shift subconscious desires is fear. Fear of insecurity, rejection, loss, grief or age.

In the case of political advertisements, the industry uses fear of the opponent candidate as the primary tool to dissuade voters.

Emory University psychologist Drew Wenston told CNN that: "fear-based attack ads are effective because they tap into a voter's subconscious." Although people say that they are not influenced by negative advertisements when asked, tests reveal that they are *subconsciously* influenced. Their brains have reacted to the ad, whether they consciously recognise it or not.

*The group watched <u>Hillary Clinton's</u> "3 a.m."
campaign ad, which was intended to make
voters question <u>Barack Obama's</u> experience.
Viewers said that the ad was fear-mongering
and that it did not make them think Clinton was
a stronger leader than Obama. But the data,
Westen said, showed that their brains reacted
differently.*

*Voters had the greatest hesitation with words
like "weak" and "lightweight" during the color
test. Westen said this meant the ad made them
question Obama's readiness.*

...

*This happens because the ads trigger a re-
sponse in the part of the brain called the amyg-
dala, which experiences emotions such as fear.
When it is aroused, it overrides logic, according
to Westen.*

In the same way, ads about aging or health or hy-
giene are aimed squarely at provoking fears, and thus
subconsciously increasing the desire of the remedy to
that fear: anti-aging cream, doctor visits, Lynx deodo-

rant and so on. A lot of the time we do not know that these fears are being provoked.

In conclusion, there is a risk that the advertisement industry undermines free will by targetting and manipulating our subconscious desires and motivations, which then influence and shape our conscious (or instinctual, reflexive and irrational) decision-making regarding the purchase of products.

Potential Solutions:

The best thing we can do is to educate ourselves on the tools that advertisers use, so that we can consciously recognise when advertisements are attempting to influence our decision-making (lol), and to understand the rhetorical, psychological, and neuro-marketing tools at the advertiser's disposal.

---

[1] Pamela Odih, *Advertising in Modern and Post-modern Times* (Sage, 2007) 10-15.

Chapter 7:

Modern Life: A Consciously Celebrated Materialism

The anthropologist David Graeber has this idea that culture has been relabelled as consumption. So: what were formerly the traits of the hippies, the principles: free love, free expression, peace not war and so on, have been replaced by the traits of the hipsters: buying iPhones, buying jeans, buying into tech start-ups and technological industries.

A consciously celebrated materialism replaces the internally conscious celebration of ideas, so that ideas themselves become an endangered species. Universities move from challenging thought, to reinforcing existing work practices – asking questions like "what do employers want?", "what do employers need?", "how do we give students workplace skills?" instead of questions like, "how do we make university students answer the biggest questions of our time?", "what is the meaning of life?".

In other words, the underground up-swell of *principled* rebellion becomes morphed into a consumeristic rebellion – a rebellion or call to arms against, yet alongside major brands, to move over and buy other, less major brands, which soon become major brands in their own right. Behavioural attitudes and dress codes at work become the measure by which people target their non-conformity, so now beards and moustaches are more acceptable in corporate environments, but so long as you never complain about company practices on twitter, or ever express anything except the correct political opinion about Anzac day celebrations.

It is okay not to conform, encouraged even, so long as your non-conformity has nothing to do with raising questions about ethics, morality or social justice; as these are ideas that are 'incompatible' with the business' 'culture,' standardized anti-social media contracts and the continuation of hard work, not smart work, for no ends, not purposeful ends.

It is ironic that, now in particular, hard work has become cherished more than ever before when the vast majority of the population are working in service industries where hard work involves long hours of paperwork that should have been automated by computers anyways – which were meant to absolve us of paperwork but have otherwise entrenched an environment of internal boredom, futility and stagnation. The principle of the Protestant work ethic has been hijacked by a principle of hard work without ends, as in the futility of rolling a boulder up a mountain, for infinity – as in the myth of Sisyphus. What was once a Greek punishment from the Gods is now everyday life; the futility of refiling the same paperwork, reopening the same programs everyday, re-looking at the same websites. We are always expecting different outcomes, but come to be depressed by the routine and monotony and sameness of the outcomes we receive.

If you are a young person and you do not want to work a forty hour work week, the media will demonize you as part of a "fantasy entitlement/no responsibility/no consequences" generation; a generation sold on too many dreams and not enough realities. Many young people are told to work in retail, hospitality or hard labour in apprenticeships, during their studies, as a form of paying off their dues or taking on real 'adult' responsibilities.

This is despite the overwhelming amount of evidence proving that all of these industries are reckless choices with regards to personal health, finance and happiness. These metrics are not required or necessary, so much as the metric of "what contributes to the economy?"

According to the American social researcher Charles Murray, over the last fifty years blue collar workers have become more and more likely to lose their jobs, become unemployed, divorce, get sent to prison and vote less often when compared to white collar workers, and the trend is only expected to get worse.[1] Yet stories in Australia, including the recent collapse of the car manufacturing industry, are not used to warn young people away from primary industry, but to reinforce the bold myth of menial tasks *as* meaning, hard work *as* an end, labour *as* life. Attitudes of 'toughing it out' are said to be associated with the construction industry having a male suicide rate that is twice the national average. Still, we tell people to "tough it out" and join these industries to begin with. Still we tell young people to "quit complaining".

The retail and hospitality industries are not that much better when it comes to personal satisfaction. A recent *Retail Workforce Study* revealed a crisis of confidence in retail employees due to "negative perceptions of the industry," with many workers viewing retail as a "short-term job with limited career opportunities". A 2006 study from Curtin University, reveals the deeply personal, psychological issues associated with these positions – some of which are likely predeterminates for lifelong mental illnesses. Futility and a lack of meaning in our jobs causes illness and death, there, that's as plain as it can be said.

Typical answers to the survey, from actual retail and hospitality workers, included: "Nobody notices you"; "I often think what's my purpose here[?]"; 'You work in a thankless environment"; "I want to get out as soon as possible"; "they don't think we're human".

A question must be asked: why not abolish these jobs entirely?

On the one hand, young people are condemned for being materialistic, vapid, self-serving, social media maverick narcissists. On the other hand, the media cries existential crisis the minute retail figures start falling below national averages at Christmas time.

Either materialism is terrible, in which case end it, or materialism is necessary 'for the economy', in which case, vapid self-serving narcissists are going to be your citizenry. Friends, Romans, vapid consumerist narcissists, rejoice.

If you want young people to grow up and not be narcissistic and materialistic; then abolish the retail industry, or at the least, stop sending young people to go work in it.

If that's too much hard work, then get robots to do it for you.

Chapter 8:

Lifetime Employment in Japan: Casual Work, Part-Time Work and Women Under Equal Opportunity Law

Lifetime employment has long been the cornerstone of corporate governance in Japan. College graduates at large firms have traditionally been guaranteed employment until retirement. These graduates, almost exclusively men, are guaranteed job security in return for complete loyalty to their company of choice.

Originally sustained by cultural forces of "loyalty" and collectivism, the lifetime employment system faces new threats; Equal Opportunity law, women entering the workforce and flexible work practices.[1] Yet despite these threats, lifetime employment has remained relatively stable in recent years. While few employees are moving *into* the system, it is equally true that few are moving out.[2] This essay argues that the future fate of lifetime employment will therefore depend not on Equal Employment Opportunity law, or other legal measures, but rather economic and cultural factors. There are several competing economic and cultural factors that currently cement the system in place, and prevent any flexibility. This status quo is likely to continue.

Regarding women in work, I argue that the increasing role of women in secondary labour has been a mixed blessing; offering opportunities, but also entrenching discriminatory barriers. If lifetime employment comes to an end, women will have a more fulfilling role within the Japanese economy. However, the prospects of full gender equality remain an unlikelihood, and it will take a long time for Japan to reduce entrenched sexism in the workplace.

Theory:

To understand changes to lifetime employment, it is important to look at what currently cements the system in place, namely the system's cultural and legal foundations.

Culture:

The lifetime employment system was traditionally entrenched by Japanese cultural traditions of loyalty, collectivism and social harmony.[3] Employees, loyal to their company (chosen at graduation), were reluctant to leave due to that loyalty.[4]Employers, bound by social pressures to retain their male workers (as they tend to be the sole providers for households in Japan) have been reluctant to lay off permanent staff.[5] Presidents of companies have called the laying off of staff "the worst sin of an employer".[6] These cultural views have entrenched an inflexible work environment in Japan, based on cultural loyalty over economic efficiency or gender equality.

Explaining lifetime employment through culture is called the "cultural model" of academic theory.[7] Critics of this model argue that "culture assumes a set of... universal" norms and that lifetime employment is not universal in Japan.[8]However, this critique purports a very narrow definition of what "culture" actually entails. A culture need not dominate every single industry. Subcultures in employment practices are as much a reality as subcultures in fashion.[9] Just as part-time workers, dispatch workers and fixed-term employees form their own 'subcultures' in Japan, so too lifetime employees have their own subculture in the wider Japanese economy.[10] Within the "lifetime employment" subculture, values of loyalty and social harmony have a higher prevalence than within, say, "dispatch worker" communities, where the institution prevents sustained employer/employee contact.[11] In lifetime employment, long service promotes familial relationships with colleagues.[12] Companies remain loyal to employees, who remain loyal to companies.[13]

It is not surprising therefore that lifetime employment came to prominence in the 1970s onwards, during Japan's economic boom. To view culture as something outside of economic forces is irrational. Rather, the argument can be put that as Japan's economy grew, so too did the 'lifetime' subculture alongside it.[14] Lifetime employment's sub-culture proliferates in economic boom periods, where long-term employment is a sustainable practice.[15] If Japan's economy were to improve, therefore, this cultural force may once again rise in prominence.[16]

Dismissal Laws:

Lifetime employment has also been entrenched by laws on unfair dismissal of permanent staff.[17] The courts have restricted employers firing lifetime employees in an abusive manner, in an "unreasonable" manner or in a manner that goes against the "common sense of society".[18] This case law has been codified in the *Labor Contract Act* (2007) where any dismissal of permanent employees requires "reasonable grounds".[19] The term "reasonable" is a high bar.[20] Case law establishes that even if a radio presenter sleeps in and misses one and a half shifts it is "unreasonable" to dismiss them.[21] If someone looks "gloomy" in an interview, it is "unreasonable" to dismiss them.[22] Laying off workers to improve dividends is "unreasonable".[23] Laying off workers, even in economic recession, should be a "last resort".[24] The "last resort" rule forms part of four factors now used to help analyse abusive dismissals.[25] These four factors suggest, but do not conclusively demand that: dismissal should be a business necessity, a last resort, objective and reasonable, and fully consultative.[26]

The result of these decisions, the four factors and the legislation is that employers are very limited in how and when they are able to fire permanent staff. The resulting employment inflexibility bolsters "lifetime employment".[27] Companies transfer employees internally and to subsidiaries, instead of firing them.[28] Workers tend to stay working with a company (in some capacity) for life. The courts themselves argue that "efforts should be made to transfer and absorb redundant personnel" within the company, before staff are laid off.[29] Alternative measures should be tried *first*, otherwise dismissal may be deemed "unreasonable". If an employee *does* get dismissed, they will have legal recourse to these strong protections under case law and statute.[30] This results in unfair dismissal law indirectly supporting a "lifetime employment" model, where employees cannot be fired easily.

Critics of this "market" analysis argue that lifetime employment *pre-dates* unfair dismissal laws, and therefore there is no causative relationship.[31] Here, I would argue that even if lifetime employment has other *origins*, in politics or culture, this does not prevent it from being *buoyed* by unfair dismissal laws. Indeed, but for unfair dismissal laws, lifetime employment would have declined more rapidly in the 1990s and 2008-2009, during Japan's economic crises.[32] In the West, staff are dismissed during economic downturns.[33] However in Japan, instead of firing staff, firms changed workplace practices; reducing the hiring of new graduates, internally transferring employees to subsidiaries and freezing wage growth.[34] The firms who *did* lay off staff were met by strong resistance from case law, as cited above, and were urged to restructure.[35]Several dismissal decisions were overturned.[36] In this way, unfair dismissal law played a significant role in bolstering lifetime employment during the years of economic recession – forcing firms to use other methods of flexible workplace practices. Indeed, to fire all workers and start from scratch would be a legal impossibility under the current law, meaning that even if companies wanted to end the system, they would have to wait until *at least* the retirement age of all *current*lifetime employees.

In light of the above, it is not surprising that most firms (85%) still commit to lifetime employment in one way or another.[37] To do otherwise in the short term is impossible, because firing staff arbitrarily is a legal impossibility. Therefore, unfair dismissal laws support the system's continuation.

Rise of "Secondary Labour":

Those who argue that lifetime employment is in substantial decline point to the rise of secondary labour as a new threat.[38]

i) Statistics:

Whether the rise of secondary labour is a threat to lifetime employment depends on the statistical model you use to evaluate the data. If the *entire* economy is considered (Appendix 1), then lifetime employment has remained stable at around 57% of the economy from 1990 to 2001.[39] A shift has instead taken place from workers in self-employment, family work, piece work and agriculture work into fixed-term and part-time work.[40][41] From 1990 to 2001, the number of non-standard employees and part-time employees rose by 13.6%.[42] In the same period, self-employment, family work, piece work and agriculture declined by 16.6%.[43] (Unemployment rose by 2.6%, accounting for most of the discrepancy between these two figures).[44] There has, therefore, been a move away from informal work structures towards part-time and fixed-term employment.[45] Lifetime employment, as a factor of the entire economy, has only declined by a minor 0.5%.[46] More recent figures affirm that lifetime employment has remained relatively stable at two thirds of the workforce throughout the 1990s, through to 2010.[47]

When the statistical analysis is limited to *employees* (Appendix 2), *not the whole economy*, lifetime employment looks to be in a more substantial decline.[48] From 1990 to 2001, lifetime employment decreased by 6.5% as a factor of *employees in Japan*.[49] This figure excludes: self-employment, family work, piece work, agriculture and unemployment.[50] It is therefore, inflated. The decrease of lifetime employment as a percentage *of employees* is due primarily to an increase in part-time and fixed-term workers entering the workforce.[51] In absolute terms, there are more secondary workers than ever before. Recent figures show the percentage of non-regular employees rising from 20% in 1990 to 34% in 2012.[52] However, as stated above, as a percentage of *the entire economy*, lifetime employment has remained stable.[53] The lack of *growth* in lifetime employment is indicative of the impenetrability of the lifetime employment system for new workers, as opposed to an absolute decline in the system itself.[54] Companies have simply reduced the number of staff to whom lifetime employment applies by hiring new part-time workers instead of lifetime employees.[55] The amount of workers moving *out of* lifetime employment is however low, at 0.5%.[56] Hence, the lifetime employment system is simply *not grow-*

ing and *not allowing new workers to access it.* Any *decline* in the system is small. Therefore, calling the system "dead" is an overstatement. However, if companies continue to avoid hiring permanent employees, the system will continue to be in a constant, gradual, decline until its demise.

The idea that lifetime employment is on its deathbed, though suggested by several academics, may be overstated, as it assumes companies will permanently change hiring practices and stop hiring any future lifetime employees whatsoever. The evidence however, points in the other direction. When asked if they will keep lifetime employment, 78.6% of companies surveyed in 1999 said yes, 85% of companies in 2001 said yes and 79% of companies in 2002 said yes.[57] Even firms who are committed to the system however, are being hard pressed by slow growth in the economy.[58] Half of 800 firms surveyed by *Nikkei* said they can "no longer sustain permanent employment practices".[59] If the system becomes unsustainable, companies will change hiring practices.

However, the evidence of a looser commitment in hiring is not there with regard to tenure positions. Ten-year job retention rates of core male employees changed little from 1977 to 1997. [60] Retention rates increased as late as 2000.[61] This proves that, even during tough economic circumstances, businesses remain in favour of lifetime employment.

The right question might even be, "will lifetime employment have a resurgence once economic conditions [in Japan] recover?"[62] Currently, Japan's economic growth is very low, and has been since the 1990s, so it is hard to say for sure whether the lifetime employment system is in systemic decline, or temporarily affected by Japan's low growth. There is a view that the system could revitalize if the economy improves.[63] It may well be that an upswing in the Japanese economy will allow businesses to hire new graduates into the system. This trend of significant new hires (into lifetime employment) during economic booms has historical precedent, at the end of the boom years in 1991 before the collapse of the economy.[64] Hence any trend upward and downward of lifetime employment may depend upon the economic conditions in Japan. To declare lifetime employment "dying" would be premature, without knowing these future conditions.

ii) Legal Analysis:

Looking beyond economic circumstances, it is important to question *why* an increase in secondary labour has occurred over time.

Part of the answer lies in law reform, which has played a significant role in changing workplace practices.[65] Historically, Japan's legal system inadvertently entrenched discrimination against secondary labour workers, resulting in longer hours, lower pay and lower working conditions for those in the second-tier economy.[66] Recently, the Japanese government began addressing this discrimination. In 2008 the Government passed an act mandating that employers "endeavour" to provide "balanced treatment for part-time workers" and lifetime employees, even if they work different roles.[67] This Act overruled case law, whereby part-time workers were only eligible for equal pay if they worked the same "job".[68]Considering only 1.1% of businesses had part-time workers in such a job, the new Act was pivotal in widening the mandate.[69] That said, the Act has been criticised as "toothless", because it asks for an "endeavour" as opposed to an obligation.[70] However, employers *are* obliged to explain *why* part-time workers are treated differently.[71] Employers are also obliged to "endeavour" to take measures to promote part-time worker "transitions to ordinary worker status".[72]Since the reform, 48.6% of businesses have implemented transition procedures: meaning the law created effective social change.[73] Part-time workers have since gained a rare opportunity to access the lifetime employment scheme.

Fixed-term workers have received similar protections and are now given guaranteed employment until the end of their contractual term "unless there are unavoidable circumstances" requiring dismissal.[74] This wording is similar, but lighter, than the "reasonable grounds" test in the case of regular employees; but nevertheless grants a degree of job security previously unheard of.[75]

Future economic predictions suggest that as job security for non-regular employees increases, so too will job mobility, leading to an overall decline in lifetime employment.[76] Workers will be more prepared to test their worth on the open market.[77] The statistical evidence shows this trend is already occurring. There has already been an upward trend in job mobility since 2000.[78] Employees are already less satisfied in lifetime work.[79] One quarter of permanent employees express "a fading sense of belonging to their firm".[80] The greater the appeal of the secondary labour market, with its flexibility and increasingly secure foundations, the more employees will start considering shifting into it, threatening the lifetime system.

iii) Liberalization:

Another threat to the lifetime system comes in the liberalization of informal work structures, such as dispatch employment. Dispatch employment, where companies hire temp workers through third party agencies, was initially banned in Japan.[81]The 1986 repeal of this ban, along with new liberalization of the sector, directly threatened lifetime employment.[82]Originally limited to only "specialized work", dispatch workers can now work in any industry, save those listed under the *Dispatch Worker Act*.[83] This allows companies to offset the cost of hiring new graduates by hiring dispatch workers instead.

Some academics suggest that dispatch work will however remain in an ancillary role in the Japanese labour market and will never substitute "for direct hiring of permanent employees".[84] This claim is backed up by reference to the "three year" contract limitation on companies using dispatch workers.[85] However, new legislation is set to "allow companies to use dispatch workers in the same job" indefinitely.[86] This would allow companies to replace employees with permanent dispatch workers or dispatch workers on a rotating basis for an indefinite period. The risk here is that dispatch workers (who are easier to dismiss) will become a more appealing choice for employers than permanent employees.

The threat of this on lifetime employment remains small, with dispatch workers comprising "only 2.5% of all employees".[87] While their numbers more than tripled from 1999 to 2006, they still remain a very small percentage of the working popula-tion.[88] Finally, although each deregulation has led to an upsurge in dispatch workers; even full deregulation is unlikely to move a 2.5% population to quickly threaten the 67.9% population of lifetime employ-ees.[89]

Indeed, new law reform may result in the exact opposite conclusion: that dispatch workers come to facilitate a resurgence of lifetime employment. The *Workers Dispatch Act* (2012) lifts the ban on "temp-to-perm" positions, meaning that companies can now directly employ dispatch workers as lifetime employees, following the term of their dis-patch.[90] The Act also bans employers from firing lifetime staff, only to rehire them as dispatch work-ers.[91] The new law therefore promotes only one-way mobility: into, but not out of, the lifetime system.

iv) Fixed-Term Law Reform:

Similar promotional pathways into lifetime employment now exist for fixed-term workers. Article 18 of the *Labor Contract Act* 2012 mandates open-ended employment for fixed-term workers who work "continuously" for a company for over five years.[92] On the surface, this is a ground-breaking new pathway into lifetime employment. However, companies are already adjusting employment contracts to avoid the obligation: setting a three or four year contract renewal period.[93]

The new law does include protections to mitigate this kind of response. Companies who repeatedly renew a contract, or dismiss an employee who had a reasonable cause to expect a renewal, or refuse to renew the contract in a way that amounts to dismissal, will have to prove they had "justifiable cause".[94] What a "justifiable cause" is will be decided by the courts.[95] Given their hard-line approach to unfair dismissal, the courts may adopt a strict construction that reinforces fixed-term worker rights.[96] If so, it will become a lot harder for companies to *not* renew a fixed-term contract. In turn, this would make it more likely for contracts to be renewed beyond the five year term above – in which case the fixed-term employee would be offered open-ended employment, potentially propping up the lifetime employment system.

Conclusion:

A statistical analysis reveals lifetime employment has only decreased by 0.5% as a factor of the economy, and remained stable at around two thirds of the economy in recent decades. This very slow decline might well continue until the system's demise.

However, this seems unlikely. Firstly, law reform has given new avenues for part-time, fixed-term and dispatch workers to access the lifetime system. Secondly, secondary labour, far from threatening the system, may yet boost its numbers. The economic conditions of Japan may also improve, and this may give the system new life. In fact, lifetime employment will likely only face extinction is if employers change hiring practices and refuse to hire any new lifetime employees. Hence, declaring lifetime employment "dying" is premature. The system is currently stable, and may move up or down depending on future legal and economic trends.

Impact on Women?

Having concluded that lifetime employment has only slightly declined, the question remains: has there been a significant impact on women? The answer, as I discuss below, is that women still remain largely excluded from lifetime employment. As a result, women have sought more flexible jobs. The rise of secondary labour has given women new opportunities to enter an economy traditionally favouring male, lifetime employees.

-

Culture:

Japan has a long cultural tradition of women being marginalized into "childbearing and childrearing", with the result that most women have been traditionally excluded from lifetime employment.[97] Lifetime employment is simply incompatible with child raising. Systems of internal transfers without notice, tough travelling regimes, after-work social drinking and networking and other programs are prohibitive.[98] These programs create informal internal barriers that exclude women, who *must*, by cultural mandate, look after children.[99]

The exclusion of women from lifetime employment is buoyed by unequal distribution of unpaid work. As of 2006, women spend a grossly disproportionate amount of time on unpaid work in Japan, as compared to men.[100] In their 30's (often the only time women can "convert" into lifetime employment), women spend 7.5 times more hours, than men.[101]Women are therefore prevented from committing to the "absolute loyalty" demanded by the system's tough hours.[102] As a result, women are forced to consider other, more flexible jobs.

Many women in Japan are aware of these challenges, and some are refusing to marry as a result.[103] Employers, by extension, are rewarding this type of unmarried woman by allowing them into lifetime employment.[104] The number of unmarried women in their 20's has doubled from 1985 to 2005, and now rests over 60%.[105] The average age of marriage has also increased. Women are foregoing family, to secure higher paid, longer-term positions.[106] The marriage age in Japan has shifted upwards, towards an average of 26.4 in 1994, and 28 years old as of 2011.[107] However, the number of women entering lifetime employment remains very low, and what is known as the "M" curve still exists, even if it is pushed back into their mid-30s. [108]

History:

Traditionally, cultural expectations led women to drop out of the labour force after having children, and return when their child reached independent age, leading to Japan's infamous "M" curve.[109] Several large companies in the 1980's mandated that women retire upon having children, or dismissed women during childbirth.[110] Women returned after their child became independent.[111] Yet in leaving, they missed out on seniority-based wage systems.[112] In this way, the lifetime employment system entrenched gender discrimination.

Since the late 1980s, the M curve shifted positively due to a range of factors including shifts in marriage age, new secondary labour opportunities and law reform.[113] The effect on women has been substantial. Women have come to dominate secondary labour and non-regular employment. In 2007, 90% of part-time workers were women.[114] In the same year, women held the majority of positions in dispatch work (60%) and non-regular work (69%).[115] The rise of women in secondary labour can partially be explained by the marriage rates decreasing (as discussed above), but also comes down to a shift in the "M" curve, and changes to law reform.

Law Reform:

Recent law reform explains why some of the decrease to the "M" curve has occurred. As early as 1966 the Supreme Court found "compulsory retirement" of women upon marriage or childbirth illegal. [116] In 1981, the Court found that separate mandatory retirement ages were "unreasonable".[117] However, it was only after the Equal Employment Opportunity Law (1986) that companies began abolishing mandatory retirement programs that discriminated against women. [118]

Cultural reasons explain why the EEO law succeeded where case law had failed. Firstly, there is a cultural expectation in Japan for employees to resolve their differences outside of the court system through internal processes and negotiation.[119] Secondly, there are large social and emotional costs involved in taking an employer to court.[120] When it became clear that companies were not responding to case law, the Japanese government was forced to enact statutory reform. Parkinson calls this the use of "law as an instrument of coercion to bring about" a "desired social goal".[121] The EEO law banned employers from using "marriage, pregnancy or childbirth" to justify early retirement ages or dismissals of female workers.[122]A similar provision was legislated in 2006.[123] The "M" curve trend of firing pregnant women became illegal.

The effect of the EEO statute on company practices and female labour force participation was dramatic. After the EEO law came into effect, many companies took steps "to remove unfavourable terms/conditions applied to only female applicants", to remove mandatory retirement age discrimination, and to change advertising practices to advertise for both genders.[124] Companies began shifting towards equal pay for graduates of both sexes. The rate of equal pay rose from 36% in 1980 to 79% in 1987.[125]

However, instead of granting women access to the lifetime employment system, 60% (of 148 firms surveyed) introduced a new two-track system, right after the EEO law came into effect.[126] Now almost all firms operate a two-tier system, where men go into the permanent managerial track while women go into a less secure, lower paid secretarial track.[127] In this way companies avoid equal pay requirements under the law, by simply assigning women different jobs to men.[128]Here, EEO law reform was ineffective, supplying formal equality, but in practice resulting in women working entirely different jobs.[129] In 1990, 99% of men were employed in "Track A", the management track, as compared to 3.7% of women.[130] By comparison, 96.3% of women were employed in "track B", the secretarial track, as compared to 1% of men.[131] The figures have changed little since then.[132] Women remain a scarcity in lifetime employment and are hence firmly relegated to secondary labour.

Despite the EEO law's ineffectiveness in granting women access to lifetime employment, it still had a dramatic effect on increasing female labour force participation. Women in unprecedented numbers began accessing secondary labour after the law passed. Female labour force participation rose 78% from 1985 to 1997.[133] 60% of this was in the secondary labour force.[134] The number of female part-time workers also increased by 10.9% in the 13 years following the law, as a ratio of the total number of women in the economy.[135] These figures would suggest that the EEO law gave women new confidence in entering formal working arrangements (other than lifetime work). Women moved from family work, piece work and agricultural work into the more formal work structures of part-time and dispatch work over the period.[136] Firms in the late 1980s expressed a preference for hiring part-time female workers, because of their flexible conditions and lower pay.[137] Discrimination in the law here acted as an ironic *benefit* to women seeking part-time work.

The EEO law did *not* increase the hiring of married or single women into the lifetime employment system.[138] Only women under 40 with tertiary qualifications got greater access to the system, and even then, access was limited.[139]Labour law reform entrenched, rather than minimised, gender divisions regarding permanent employment.

Other Law Reform:

The EEO law was assisted by other anti-discriminatory laws allowing women a more secure place in secondary labour. Changes to the *Labor Standards Act* in 1997 for instance, made it easier for women to access flexible jobs. Article 19 banned employers from dismissing women during a "period of absence before and after childbirth" without compensation.[140] Time for child care was mandated for an hour a day.[141] Women were given new opportunities to "have it all", in terms of having a child *and* flexible workplaces. The result of these laws, along with a slow shift in culture, has seen Japanese women gain a steadily increasing foothold in secondary labour, with female labour force participation rising steadily from 2000 till 2010.[142]

The longer term historical trends paint an even more dramatic picture. In 1975, only 49% of women were employed in Japan's workforce, in 2009 this rose to 60%.[143] In 1990, part-time female workers constituted only 28% of all female workers, in 2009 this figure rose to 43%.[144] The increasing prominence of females in part-time and secondary labour work can partially be put down to law reform that mandated equal employment opportunities.[145]

Having said all of the above, Japan's female employment rate of 60% is still lower than other developed countries including: Germany at 64%, the US at 66% and Norway at 75%.[146] Women remain underemployed, since they are –at large- barred from entering lifetime employment and are forced to choose more flexible jobs.[147]

Conclusion:

Women now play a crucial role in secondary labour in Japan. Equal Employment Opportunity laws have cemented their role over time, offering new, flexible, part-time jobs. Few have entered lifetime employment, but many are balancing familial commitments with more flexible work arrangements.

As established above, lifetime employment will only decline if economic circumstances or hiring patterns of employers change. While such a case is merely a hypothetical, if the system did collapse, women could gain a foothold in more long-term employment. As it currently stands, such an outcome is unlikely, for the lifetime employment system is only in a slow decline and is not 'on its deathbed' as many academics suggest.

Appendix:

Appendix 1.[148]

Appendix 2.[149]

Bibliography:

Books/Academic Journals/Online Sources:

- Abe, Yukiko 'Long-Term Impacts of the Equal Employment Opportunity Act in Japan' (2013) 10 *Japanese Law Review* 2.

- Adams, Jonathon *Temp Nation: The demise of 'lifetime employment in Japan* (May 18 2010) Global Post < http://www.globalpost.com/dispatch/comm erce/100510/japan-economy-temporary-workers>.

- *Amendments to the Worker Dispatch Act for Employers Using Temporary Workers*, Client Alert, Baker & McKenzie (2012) < http://www.bakermckenzie.com/files/Public ation/4cb11a36-f12a-491d-837e-c03c1c1d6eab/Presentation/PublicationAttac hment/4c6c9a39-605a-4258-afa8-c327781084d8/al_tokyo_amendmentsworker dispatchact_sep12.pdf>.

- Araki, Takashi 'The Widening Gap Between Standard and Non-Standard Employees and the Role of Labor Law in Japan' (2011) 8 *University of Tokyo Law Review* 4.

• Ariga, Kenn, Giorgio Brunello and Yasushi Okhusa, *Internal Labor Markets in Japan* (Cambridge University Press, 2000)

• 'Asia's Lonely Hearts: Women are Rejecting Marriage in Asia. The Social Implications are Serious' (August 20, 2011) *The Economist* (Print Edition).

• Barrett, Kelly 'Women in the Workplace: Sexual Discrimination in Japan', 11 *Human Rights Brief 2.*

• Cole, Elizabeth, L. Mark Weeks and Yumiko Ohta, 'Amendments to fixed-term labor contracts in Japan' (Feb 13, 2013) *Orrick Herrington & Sutcliffe LLP.*

• Edwards, Linda N 'The Status of Women in Japan: Has the Equal Employment Opportunity Law Made a Difference?' (1992) 71 *Center on Japanese Economy and Business Working Papers*, 6.

• 'Employment Status Survey' (2007) *Ministry of Internal Affairs and Communications, Japan.*

• Fujii, Yasuhiro and Kana Itabashi, *Amendment to the Labor Contract Act*

(Protection of Contract Employees), Baker & MacKenzie Client Legal Alert (August, 2012) <http://www.bakermckenzie.com/files/Publication/f9010182-bd79-471a-91c7-cd5c5be8e110/Presentation/PublicationAttachment/1c0e794c-a75f-49a0-b9fd-ea8c0e9ad567/al_tokyo_amendmentlaborcontractact_aug12.pdf>;

• Gordon, Bill *Equal Employment Opportunity Law System and Women* (1998) 2 < http://www.bill-gordon.net/papers/eeol.htm>;

• Hamabe, Yoichiro 'Inadvertent Support of Traditional Employment Practices: Impediments to the Internationalization of Japanese Employment Law' (1993) 12 *UCLA Pacific Basin Law Journal* 306.

• *Japan Employment Law Update* (April 2013), Herbert Smith Freehills < http://www.herbertsmithfreehills.com/-/media/Files/PDFs/2013/Japan%20Employment%20Law%20Update%20-E-%20April%202013.pdf>.

• 'Japan Worker Dispatch Act Passes' (June 26, 2012) *International Operations Update, Radius World Growth Experts*.

- Knapp, Kiyoko Kamio 'Don't Awaken the Sleeping Child: Japan's Gender Equality Law and the Rhetoric of Gradualism' (1988) 8 *Columbia Journal of Gender and the Law* 143.

- 'Labour Force Survey' (2013) *Ministry of Internal Affairs and Communications*.

- Lam, Alice 'Equal Employment Opportunities for Japanes Women', in Janet Hunter (ed.), *Japanese Women Working* (Routledge, 1993).

- 'Marriage', *The Ministry of Justice, Japan* cited in *Marriage in Japan*, All in Japan: Information About Culture of Japan (2011) < http://www.allinjapan.org/marriage-in-japan/>.

- McAlinn, Gerard (ed), *Japanese Business Law* (Kluwer Law International, 2007)

- Meyer-Ohle, Hendrik *Japanese Workplaces in Transition* (Palgrave Macmillian, 2009).

- Milhaupt,. Curtis J 'On the (Fleeting) Existence of the Main Bank System and Other Japanese

- Economic Institutions' (2002) 27 *Law and Social Inquiry* 425.

- Miller, Robbi Louise 'The Quiet Revolution: Japanese Women Working Around the Law' (2003) 26 *Harvard Women's Law Journal* 163

- Moriguchi, Chiaki and Hiroshi Ono, 'Japanese Lifetime Employment: A Century's Perspective' in Magnus Blomstrom and Sumner La Croix (eds), *Institutional Change in Japan: Why it Happens, Why It Doesn't* (Routledge, 2006).

- Mouer, Ross and Hirosuke Kawanishi, *A Sociology of Work in Japan* (Cambridge University Press, 2005).

- Oda, Hiroshi *Japanese Law* (Oxford University Press, 3rd ed, 2011).

- *OECD Economic Surveys Japan* (April 2013) Organisation for Economic Co-operation and Development (OECD) <http://www.oecd.org/eco/surveys/Overview%20Japan%202013%20English.pdf>.

- Okunuki, Hifumi 'Labor Law Reform Rasises Rather than Relieves Worker's Worries' (2013) *The Japan Times* 1.

- Onodera, R 'Arbeitsverhaeltnisse in Japan' in P Hanau et al. (eds), *Die Arbeitswelt in Japan und in der Bundesrepublik Deutsch-land: ein Vergleich* (Cologne, 1984)

- Parkinson, Loraine 'Japan's Equal Employment Opportunity Law: An Alternative Approach to Social Change' (1989) 89 *Columbia Law Review* 604.

- Picard, Robert R and Groth, John C 'Japan's Journey to the Future' (2001) 39 *Management Decision* 315.

- Rebick, Marcus *The Japanese Employment System Adapting to a New Economic Environment* (Oxford University Press, 2005).

- Sakikawa, Takashi *Transforming Japanese Workplaces* (Palgrave Macmillian, 2012).

- Shire, Karen A. 'Stability and Change in Japanese Employment Institutions: The Case of Temporary Work' (2002) 84 *ASIEN* 21.

- Sōmuchō Tōkei-kyoku, (2008) *Statistics Bureau, Management and Coordination Agency, Statistics Japan.*

- 'Survey on Time Use and Leisure Activities' (2006), *Ministry of Internal Affairs and Communications, Japan.*

• Statistics Bureau, *Labour Force Unemployment Rate, Historical Data 1* (2014), Ministry of Internal Affairs and Communication Japan < http://www.stat.go.jp/english/data/roudou/>.

• *The Actual Status of Non-Regular Employment and Related Policy Challenges – Focusing Primarily on Non-Regular Employment, Career Development and Equal Treatment* (2012/2013) Labor Situation in Japan and Its Analysis: Detailed Exposition < http://www.jil.go.jp/english/lsj/detailed/2012-2013/chapter3.pdf>.

• *The Current Status and the Challenges of Dispatched Work in Japan* (2012), 13 <http://www.jil.go.jp/english/lsj/detailed/2011-2012/chapter2.pdf>

• Trice, Harrison Miller *Occupational Subcultures in the Workplace* (Cornell University Press, 1993).

• Tsuru, Tsuyoshi "Roshi Kankei no Non-yunion-ka" (Nonunionization of labor-management relations) (2002) cited in Chiaki Moriguchi, Hiroshi Ono, 'Japanese Lifetime Employment: A Century's Perspective' in

94

Magnus Blomstrom and Sumner La Croix (eds), *Institutional Change in Japan: Why it Happens, Why It Doesn't* (Routledge, 2006).

• Ramseyer, J Mark 'The Reluctant Litigant Revised: Rationality and Disputes in Japan' (1988) 14 *Journal of Japanese Studies* 1.

• Steinberg, Chad and Masato Nakane, *Can Women Save Japan?* IMF Working Paper/12/248 (2012).

• Uzama, Austin 'A Critique of Lifetime Employment in Japan (*Shushinkoyou*)' (2008) 17 *Ritsumeikin Journal of Asia Pacific Studies* 71.

• Wolff, Leon, 'The Death of Lifelong Employment in Japan?' in Luke Nottage, Leon Wolff, Kent Anderson (eds), *Corporate Governance in the 21st Century: Japan's Gradual Transformation* (Edward Elger Publishing, 2008).

• 'Womenomics 3.0: The Time is Now' (October 1 2010) *Japan: Portfolio Strategy, Goldman Sachs.*

Cases:

- Judgment of Nagano District Court, Ueda Division, 15 March 1996, *Hanta* 905-276 (*Marukō Alarm Case*).
- Judgment of the Niigata District Court, 26 August 1966, *Rominshu* 17-4, 996.
- Judgment of the Okayama District Court, 31 July 1979, *Rōhan* 326-44 (*Sumitomo Heavy Industries* case).
- Judgment of the Osaka District Court, 10 December 1971, *Rōmin* 22-6-1163 (*Mistui Shipbuilding Case*).
- Judgment of the Supreme Court, 25 April 1975, *Minshū* 29-4-456 (*Nigon Shokuen Siezō* case)
- Judgement of the Supreme Court, 31 January 1977, *Saikōsai-saibanshū* 120-23 (*Kōchi Broadcasting Co.* case).
- Judgment of the Supreme Court, 20 July 1979, *Minshū* 582.
- Judgment of the Supreme Court, 24 March 1981, *Minshu* 35-2-300 (*Nissan Motors* case).

- Judgment of the Tokyo District Court, 20 December 1966, *Rōmin* 17-6-1408 (*Sumitomo Cement* case)
- Judgment of Tokyo High Court, 29 October 1979, *Rōmin* 30-5-1002 (*Tōyō Sanso* case)

Legislation:

• *Act on Improvement, etc. of Employment Management for Part-Time Workers* (2008) [trans, *The Actual Status of Non-Regular Employment and Related Policy Challenges – Focusing Primarily on Non-Regular Employment, Career Development and Equal Treatment* (2012/2013) Labor Situation in Japan and Its Analysis: Detailed Exposition.]

• *Act on Securing, Etc. of Equal Opportunity and Treatment between Men and Women in Employment* (2006).

• *Act for Securing the Proper Operation of Worker Dispatching Undertakings and Improved Working Conditions for Dispatched Workers* (1986).

• *Employment Security Act* (1947), (Unofficial Translation, 2007).

• *Equal Employment Opportunity Law* (1986).

• *Labor Contract Act* (2007) [*The Japan Institute for Labour Policy and Training* trans (2008)].

• *Labor Contract Act* (2012) [Hifumi Oku-nuki trans, 'Labor Law Reform Rasises Rather than Relieves Worker's Worries' (2013) *The Japan Times* 1].

• *Labor Standards Act* (1997 Amendment).

Chapter 9:

20 Years on From Gangland: We've Still Got a Youth Culture in Crisis

In 1997, *Gangland: cultural elites and the new generationalism* dominated Australia's book scene. Arguing that young people were under-represented in Australia's mainstream media, subjected to 'moral panics' and increasingly demonised by the press, the book painted a picture of youth culture in crisis.

Twenty years on, very little has changed. Young people are still under-represented in Australia's media, still demonised by the press and still in a state of perpetual crisis. Then, as now, the primary culprits for the problem are Baby Boomers. As *Gangland's*author Mark Davis wrote:

> *Has an older generation of cultural apparatchiks, used to being at the centre and having a strong media presence, more or less systematically set out to discredit young people and their ideas, even progressive opinion generally?*

The answer, then and now, is yes. Having wrecked the global economy in 2008, devastated the environment, massively cut university funding, created a housing bubble and started several of the longest wars in history, Baby Boomers are capping it all off by blaming their children for their problems.

Most commonly they do so by calling the young entitled, self-serving and lazy. If houses are too expensive, a Baby Boomer millionaire suggests, then young people should spend less on avocado toast. If jobs are increasingly unstable, a Baby Boomer journalist writes, then young people should have a better work ethic.

The easiest way to avoid responsibility, as the Baby Boomers so frequently do in this manner, is to blame those suffering for the cause of their own suffering. Victim-blaming the poorest and youngest among them, in a manner that passes the buck most eloquently.

In 1997, they were doing the same thing, ignoring the problems of youth with broad generalisations, name-calling and hand-wringing:

> *Behind the hype and trivialisation that has accompanied the much-loathed moniker 'Generation X', young people are suffering. They have the highest suicide rates in the country. They are most likely to be long-*

*term unemployed. The numbers of home-
less young people have risen rapidly. They
have been among the main losers in cuts to
government services.*

Mark Davis could be writing the same book today.
Despite the moniker 'millenial' and its association of
entitlement and laziness, young people are suffering.
Today, they are dealing with the worst youth unemployment in history. They are paying three to eleven
times more for a home. They are earning 20% less
than Baby Boomers at the same stage in life. They
are working flexible jobs, with far less stability. They
are set to retire much later. They are the main losers
in cuts to government services.

We would all know this, of course, were it not for
the Baby Boomers monopolising the mainstream media, and thereby preventing young voices from speak-
ing out about their suffering. As in 1997, the young
today are massively under-represented in the mainstream media, on television and on radio. Instead,
they fill backroom positions, as anonymous ghosts
working on 'tech-savvy' projects like social media
management, video editing and sound design.

There are a few notable exceptions, such as the young celebrities who cut through the Boomer ceilings, often by being funny. As I have written before, there is a reason why young comedians are allowed a space on television: they are there literally to not take themselves seriously.

Writing for *The Conversation*, Jay Thompson argued that young people are represented more than ever before in 'traditional' media. He pointed to names like 'Clementine Ford, Josh Thomas, Nazeem Hussain, Jessica Mauboy, Hunter Page-Lochard and Benjamin Law'. Three of the examples mentioned are comedians. If the list were expanded, most of the 'millenial celebrities' would fall into the comedian category.

With a stranglehold on traditional media, it is important for Baby Boomers to limit and discredit any competition – one way to do this is to laugh at anything a young person says. One way to do *this* is to only hire young comedians.

Another way is to praise 'new media' as the voice of the young, while never listening to that voice. Social media, YouTube and other sources are viewed as democratic vehicles by which young people can access a platform. These platforms are portrayed as equal to the established traditional media networks.

Malcolm Turnbull has made the argument before that 'new media' brings about a diversity of voices and increases who is able to speak.

While true, comments like these are broadly misleading. Traditional mediums like television and newspapers are in some ways 'fixed'. There are millions of people who watch the nightly news, meaning that anyone who appears on it receives millions of 'views'. By contrast, the net is a place of warring sources of information, vying for readership, often on a much smaller scale, of the tens or hundreds of thousands.

To say that a young person with access to a blog is equivalent to a Baby Boomer with access to a television network is like saying your neighbour with a home-stitched flag is as powerful as the prime minister.

Major media companies continue to exclude young voices, and increasingly the youth are forced to carve out a niche on the newer, lesser-viewed platforms. Many are turning to YouTube, a platform that can in some ways be described as more 'democratic' than traditional media.

Allowing anyone a voice and a chance to publish, YouTube represents a failed vision of what media was meant to be about. On it, the young are showing the kind of talent and intellectual curiosity that Mark Davis described in *Gangland*. The kind of insightful, informed content that television executives seem to think is impossible to get out of a twenty-something.

Perhaps the final and most persistent myth perpetuated by the Boomer Generation is the idea that all of this – the disadvantage, the suffering, the exclusion from mainstream media – was just as bad for them 'back in their day'. The failure of Baby Boomers to understand that things change – economics, social structures, political systems – is perhaps key to their blind-spot on the younger generation. Instead of acknowledging the difficulty of the youth today and working with the young to solve critical worldwide problems, they turn their backs. Instead of empowering the young to speak out on traditional media platforms, they create more and more layers of 'expertise' requirements, certification requirements and barriers to entry.

Increasingly, the young are expected to serve them coffee for less pay, work in the backroom on anonymous projects, and build their brands on social media – to entrench their existing platforms even further.

Writing recently in *The Guardian*, Julianne Schultz had a moment of Boomer self-realisation, stating that 'the world we have bequeathed to our children feels darker than the one I knew'. In this rare moment, she can be joined by the insightful commentary of the German philosopher Walter Benjamin, who had similar insights in his own time. In closing, I will paraphrase his call for his generation to stand up and fight for something more out of their lives: 'Past generations didn't struggle so we could have dreams of a better life. They struggled so that for us a better life could be a material reality.'

This chapter was originally published on Overland.

Chapter 10:

Degrees of Separation: Companies Shed Degree Requirements to Promote Merit Over Qualifications

At the end of 2016, the Australian Bureau of Statistics
(ABS) revealed that close to two-thirds of all Australians had completed a degree or apprenticeship.

The growth in the number of people attending a
university or TAFE has risen out of a cyclical demand-
driven system called "academic inflation".

Think supply and demand. If an employer can hire
someone with a degree or someone without, they'll
hire the person with a degree because they are seen
as the superior candidate. This puts pressure on
everyone to get degrees. But once everyone has one,
the value of having a degree goes down.

A couple of decades ago, a high school diploma
was sufficient to get a job in journalism or business.
Now a bachelor's degree is required.

Where a bachelor's degree was sufficient to get a
job in research, now a master's degree is required.
Where a master's degree was sufficient to get a job in
university tutoring, now a PhD is required.

The number of people gaining master's degrees
has doubled from the early 1980s to the late 2000s.
The PhD, once a niche qualification for the few, has

become the definitive qualification of what it means to be an expert today.

For a young millennial, this means they might have to study three to six years longer than their parents did to get the same job. That's three to six years of debt, without any increase in wages at the end of it. If that job requires only basic skills such as photocopying or research, then the millennial will not necessarily have any greater skills than their parents had with a high school diploma in the 1970s.

This is borne out by data from the US, showing that in 1973 only 28% of jobs required a degree, compared to 59% in 2008.

When companies demand more and more degrees for very basic jobs, they cut off access for unskilled workers to break into the job market. In many professions, the traditional route into a job was through an apprenticeship, which required no professional training or degree, no private tuition or cost on behalf of the student.

Even a degree like law, which is today seen as a prestigious qualification, used to be taught exclusively by students apprenticing in legal offices. The cost was borne out by the business, rather than the student, meaning students could come from a variety of backgrounds, including former convicts.

As a result, universities are increasingly being blamed for cementing privilege, by entrenching the positions of the wealthy in the job market, as those with the most access to degrees.

In response to concern over diversity and equality of access and opportunity, top firms including Ernst & Young and Penguin Random House have recently abandoned degree requirements altogether.

Ernst & Young got rid of all degree requirements in 2015, explaining that a candidate's degree had no correlation to their future job performance. A year later, Penguin Random House followed suit, citing the need to hire applicants from more diverse backgrounds.

PriceWaterhouseCoopers, Ogilvy Group, Apple and Google have all relaxed their degree requirements in recent years, lowering required grades or targeting poor performing and non-college students. The idea is to hire people based on merit, rather than credentials, often by assessing candidates with psychometric testing or other performance based tests.

Instead of abandoning degree requirements altogether, some firms, including professional services firm Deloitte, have chosen to hide which university an applicant graduated from. The aim is to limit the "prestige" associated with an institution, so as to more accurately test the abilities of the applicant.

What matters in both this and the broader debate is a refocus on ability over credentials.

That some companies are relaxing degree requirements raises new questions about the value of a university education. The question is whether these few companies are outliers or the forerunners of a new trend of preferencing merit over qualifications. If the trend does persist, then the job market of the future may have as little barriers to entry as the job market of the 1970s.

This chapter was originally published on The Conversation.

Afterword

Thank you very much for reading Essays in AI, an exploration of the ways in which technology will shape the future of 9-5 work.

Please take a moment to leave a review on an online retailer such as Amazon. I welcome contact and discussion with my readers.

At my website, you can contact me, sign up to my blog and follow my Twitter account:

Blog: http://newintrigue.com

Twitter: https://twitter.com/JoshKrook

Buy my other books:

Legal Education, Privatization and the Market:

"Where once, law schools produced bastions of honourable and ethical practices, today's law schools produce bland enforcers of the law, incapable of questioning the very principles that form the bedrock of their education."

Buy here: https://www.amazon.com/Legal-Education-Privatization-Market-Australian/dp/1530801281

Us vs Them: A Case for Social Empathy:

"The modern city is a place of social circles; clusters of contacts who know each other and strangers who don't. It is a place where diverse relationships are in decline. In the city, strangers seldom meet beyond daily functions. Instead they brush by with a haste that so defines a century of 'too little time'."

Buy here: https://www.amazon.com/Us-Them-Case-Social-Empathy-ebook/dp/B00LDMSIVC